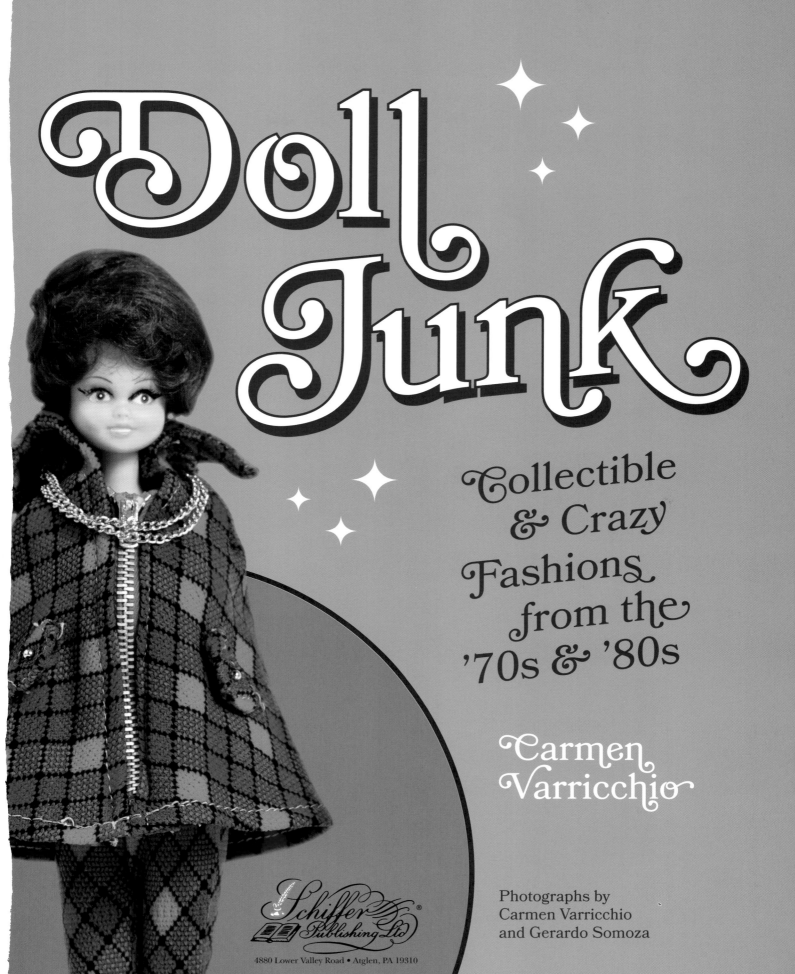

Other Schiffer Books by the Author:

Collectible Doll Fashions: 1970s, ISBN 978-0-7643-1806-1, $29.95

Other Schiffer Books on Related Subjects:

Cher™ Doll & Her Celebrity Friends: With Fashions by Bob Mackie by Sandra "Johnsie" Bryan, ISBN 978-0-7643-1970-9, $19.95

Coffee with Barbie® Doll by Sandra "Johnsie" Bryan, ISBN 978-0-7643-0412-5, $19.95

Copyright © 2015 by Carmen Varricchio

Library of Congress Control Number: 2014959833

All rights reserved. No part of this work may be reproduced or used in any form or by any means—graphic, electronic, or mechanical, including photocopying or information storage and retrieval systems—without written permission from the publisher.

The scanning, uploading, and distribution of this book or any part thereof via the Internet or via any other means without the permission of the publisher is illegal and punishable by law. Please purchase only authorized editions and do not participate in or encourage the electronic piracy of copyrighted materials. "Schiffer," "Schiffer Publishing, Ltd. & Design," and the "Design of pen and inkwell" are registered trademarks of Schiffer Publishing, Ltd.

All items shown in this book are from the collection of the author. This book is not sponsored, endorsed or otherwise affiliated with any of the companies whose products are represented herein. "Barbie" and all other Mattel doll names are registered trademarks of Mattel, Inc. The former Ideal Toy Company trademark "Crissy" is owned by Mattel, Inc. None of the toy companies mentioned in these pages authorized this book, nor furnished or approved of any of the information contained therein. Most of the items in this book may be covered by various copyrights, trademarks and logotypes. Their use herein is for identification purposes only. All rights are reserved by their respective owners.

Inside back cover image: "Thoughtprint #3 (Portrait of Carmen Varricchio)" by Ennid Berger.

Designed by Molly Shields
Cover designed by Matthew Goodman
Type set in Bookmania/Times New Roman

ISBN: 978-0-7643-4812-9
Printed in China

Published by Schiffer Publishing, Ltd.
4880 Lower Valley Road
Atglen, PA 19310
Phone: (610) 593-1777; Fax: (610) 593-2002
E-mail: Info@schifferbooks.com

For our complete selection of fine books on this and related subjects, please visit our website at www.schifferbooks.com. You may also write for a free catalog.

This book may be purchased from the publisher. Please try your bookstore first.

We are always looking for people to write books on new and related subjects. If you have an idea for a book, please contact us at proposals@schifferbooks.com.

Schiffer Publishing's titles are available at special discounts for bulk purchases for sales promotions or premiums. Special editions, including personalized covers, corporate imprints, and excerpts can be created in large quantities for special needs. For more information, contact the publisher.

DEDICATION

This book is dedicated to my parents,
Tony and Nina.

Contents

Acknowledgments 4
Introduction 5

Chapter One: Small Doll Junk 6
Chapter Two: Barbie-size Junk 20
Chapter Three: Ken-size Junk 140
Chapter Four: Big Doll Junk 160
Chapter Five: Leftovers 172

Acknowledgments

Many thanks go to my sister Roseanna and niece Gabriella for their invaluable assistance in preparing this book. I must thank my dear friend in Germany, the author and doll collector Astrid Geskes, for her great kindness and much needed help. I would also like to thank Ruth Keessen of *Doll Reader* magazine and Pat Henry of *Fashion Doll Quarterly* for allowing me to reprint material originally submitted to those magazines. Thanks to Gerardo Somoza for his excellent photography, to the artist Ennid Berger for her wonderful portrait, and to the artists Ilene Vultaggio and Joyce Kubat for their help and kind words of encouragement.

Introduction

"EEEWWW. Fake Barbie clothes." Those four words verbalize the faint discomfort my sister and I would share when, as kids "playing Barbies," we sometimes unearthed an inferior, generic, or "clone" dress or top from her sizable supply of perfect Mattel doll outfits. (My sister remembers in particular the molded hard plastic purses with excess plastic still stuck to the handles.) To this day neither of us can recall the exact provenance of these cut-rate copycats (a birthday or Christmas present from some penny-pinching relative?) as our allowance dollars were saved exclusively for Mattel outfits, and our parents simply knew better.

My sis, ever efficient, was quick to classify. The cheapies were unfailingly treated as tainted outcasts, and as such were never kept with the Mattel-made clothes in the corresponding Mattel cases. They were packed just as neatly, but in household "containers" (such as my aunt's discarded Chandler's French Room Originals shoeboxes) and left to rot.

Umpteen years later, I've softened a bit. As a doll clothes collector who has more or less had his fill of Barbie, I've turned to the very items I detested as a snot-nosed, fashion-know-it-all kid. If my first book (*Collectible Doll Fashions: 1970s*) dutifully displays some of the "big guns" of seventies doll fashion, then this new volume gets unashamedly dirty with a healthy dose of the fabulous fakes, wilted wannabes, and unmitigated junk no self-respecting doll lover would be caught taking a second look at. Here are oodles of not-exactly-gorgeous getups and some of the downgraded dolls who wore them, from the mid-1960s to about 2000, most shown in their original packaging. Prepare to shield your eyes from clumsily drawn fashion figures, brave yet pathetic attempts at high fashion lingo (you'd be surprised how many times "high" is spelled "hi"), and alternately embarrassing/intriguing package graphics. I guess with age comes a little forgiveness: call me mushy, but isn't there something inherently touching in the blatant knockoff or shapeless doll sack that no one wants?

The magic of mediocrity knows no boundaries, so I have included fashions and/or dolls from Australia, Canada, England, Germany, Italy, Mexico, The Netherlands, Portugal, Spain, and Thailand, as well. I am by no means the only collector inexplicably drawn to fearlessly featherweight dolls and clothes. Thanks to a new, younger generation of leap-for-cheap fashionistas, anonymous copies (some dropped into plastic bags with stapled-on cardboard headers) are now swathed in status as "fashion clones" on eBay, and appear to have a significant following.

This introduction, with some changes, was printed as an article in the May 2012 issue of *Doll Collector* magazine and is used with the permission of Scott Publications.

> "I LOVE my doll junk!"
> —Overheard at a New Jersey collectible doll show some years ago

Small Doll Junk

CHAPTER ONE

This Dawn-size UN Petite doll has an even higher forehead than Dawn and super-extended lower lashes. She looks like a space alien in drag.

Chapter One: Small Doll Junk   7

A nameless, raven-haired Dawn clone from Germany is packaged with a Crissy-size dust pan, brush, broom, and pail.

Kmart copied Matchbox's Suky and Patty packaging for their 6-inch Tracy dolls. Each figure was dressed to coordinate with the backdrop inside the box. Here, one doll is dressed as a nurse, administering to a patient in a hospital. The other takes a moonlit walk on a winter evening.

A Tracy gift set comes with two additional outfits and lots of arrows on the box. You'll see that motif often in this volume.

Uneeda's Tiny Penelope, with her yellow oven mitt hands and calico dress, is an obvious clone of Mattel's mid-seventies Honey Hill Bunch dolls.

Two rare Lisa clone outfits meant for Hasbro's 1971 9-inch World of Love dolls.

Chapter One: Small Doll Junk   9

This JCPenney Dawn-size outfit includes what must be the tiniest poncho EVER!

A fringed "buckskin" set for Susan. The front and back of the box uses the same illustrations found on a Shillman Mini-Mod package (the skirt on the upper left figure shown on the back was lengthened to include a midi in the group). The same sketches are copied on a German Jeany box (above).

This rare five-outfit set "For Evel And Other 7" Figures" was available exclusively through the Montgomery Ward catalog. The photo on the box shows a row of five 1972 Evel Knievel dolls dressed in the fashions. (I am guessing the full name could not be used for copyright reasons.)

These Mattel Big Jim samples include a bright Clint Eastwood poncho that to my knowledge was never used for a Big Jim doll.

Chapter One: Small Doll Junk

Here is the African-American version of Woolworth's Wee Three dolls in their original box, just one of many Mattel Sunshine Family clones. I've seen these dolls dressed in at least three different outfits.

The package design for the fashions frames the window of the Wee Three "home" with shutters, a brick wall, and even a shingled roof. Mama cradles Baby (baby?) in a hideous floral dress, apron, and bedroom slippers. Here is most of the series, with a couple of variations.

12 Doll Junk: Collectible and Crazy Fashions from the '70s & '80s

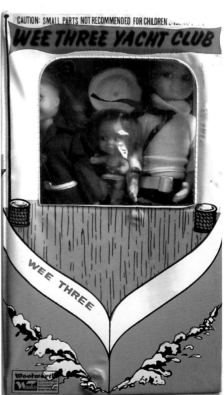

The Wee Three family, dressed for a day of yachting, nestles inside a white vinyl case designed to suggest a boat cutting through the open seas.

Chapter One: Small Doll Junk   13

A 1976 magazine ad claims the "back to nature" Mattel Sunshine Family dolls are "neighborly folks with homey family activities for your child to share." These dolls, dressed in sweet prairie dresses and denim, made pottery, spun yarn, and sold homemade crafts from a Walton-style van with a piggyback shack. Kmart's rather sad imitation, the Springtime Family, suggests a somewhat desperate trio on the dole, selling flowers and fruit for twenty-five cents from the back of a car. The Canadian version shown here trades the yellow vinyl case that doubled as the family's van for a cardboard house with bizarre, genetically altered sunflowers growing in the front yard.

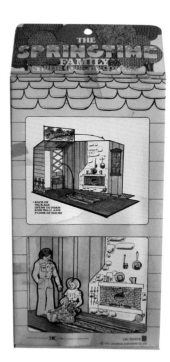

The Springtime Family has a little dachshund (The Sunshine Family had both a cat and a dog). Papa Springtime has a smarmy moustache and eyeglasses.

Doll Junk: Collectible and Crazy Fashions from the '70s & '80s

The Smart Family may be the most laughable of the Sunshine Family clones. There's something dysfunctional about this cheaply dressed group. Dad looks like he could be a wife beater.

The "Loving" Family pairs a more adult-looking mother with a father sporting a Ken-like head (Dad and Baby have matching hairdos). In a weak attempt to emulate the Sunshine Family dolls the illustrator has replaced the family's eyes with chocolate Raisinets.

Chapter One: Small Doll Junk

Mattel's Sunshine Family dolls were big sellers in Italy, prompting other American imports and Italy's own version, the Sympathy Family. The first dolls look remarkably similar to the Mattel originals. The possibly later baggie (dolls dropped into plastic bags) family shown here includes a father with an odd GI Joe–like head. Some of the outfits are repackaged Totsy Wee Three family sets (with slight variations), leading me to believe that all the clothes came from Totsy. There's lots of scaled-down seventies style here, and who can say no to a Bugs Bunny onesie (in the set at top right)?

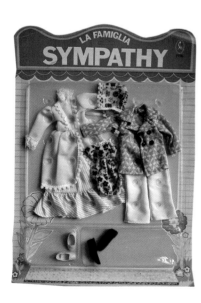

Germany's Alme Elegant fashions copy the Sympathy Family house packaging but omit any verbal suggestion of a family. Also, the sizing is incorrect—the pants are too long!

Kenner's early-seventies Jenny Jones & Baby John played on the matching mother and toddler gimmick, perpetuating a fashion fad that would reach a dizzying climax with pint-sized Pucci-esque pajamas and miniature vicuña coats. The slick and conspicuously single Ms. Jones, dressed in an expensive-looking knit pantsuit or a black Rudi Gernreich–inspired hostess gown (the doll's original outfit), had little in common with her more popular Mattel competitors, the unrelentingly wholesome, vaguely hippy-ish Sunshine Family. Jones suggests a stylish, career-conscious gal with the status and salary needed to raise a child comfortably on her own. In comparison, the Sunshine Family's ubiquitous prairie dresses, patched denim, and tie-dye seem benign and middle-class. The JJ box back shows four smart looks-with-layette called "coord-knits," three of which are shown here.

Spain's Walton-size Hogarin Family had a whopping ten members (eleven if you count the family dog): Grandpa, Grandma, Papa, Mama, The Older Sister, The Boy, The Baby, The Twins and even a Maid (the first ostensibly upper-class doll family?). Outfits, a huge home, individual furnished rooms, and furniture sets were available.

Here's a jarring sign of the times: Papa reads a miniature *Daily Mirror* with the headline "Nixon's Christmas Deluge of Death" (a reference to the infamous Operation Linebacker II bombings on North Vietnam begun on Dec. 18, 1972).

18   Doll Junk: Collectible and Crazy Fashions from the '70s & '80s

"Grandpa goes hunting"

"Grandma goes shopping"

"Grandma goes to bed"

"Mom goes on vacation"

The back of an outfit box shows the Hogarin clan in their Sunday best.

Chapter One: Small Doll Junk   19

The back of the box shows a photo image of Ms. Quant bending down to greet sketches of her two newest creations.

Most collectors are familiar with famous designer Mary Quant's Daisy doll, but did you know about Bubbles & Squeak, the toddler twosome from 1975? Bubbles had a red bob or long blond locks, and Squeak did just that when you pressed his tummy. The dolls and outfits, which were most likely available only in England, are nearly impossible to find. Here's a sweet red-corded "velveteen" dress for Bubbles called Birthday Party.

Barbie-Size Junk

CHAPTER TWO

1978 Petra Sommerwind and her new black friend Donna Sommerwind wear their original outfits.

GERMAN PLASTY PETRA, FRED, AND PEGGY DOLLS

Germany's Petra may be the best-known clone of Barbie you never heard of. Petra lasted nearly as long as "Her Pinkness" (production ended in 1993), has a devoted legion of fans across Europe, and is currently fetching high prices on eBay, but sizable gaps in the information available on her many permutations and clothing sets continue to stymie European collectors.

In 1964 the founders of the Plasty Toy Company, Helmutt Fiedler and Friedrich Podey, jumped on the Barbie bandwagon with their cheaply made but charming copy. During the first year of production, 460,000 dolls and nearly one million outfits sold, making the cut-rate cutie a certified hit. In 1966 the Petra nuclear family formed with the additions of Fred and Peggy, knock-offs of Ken and Skipper. (A 1969 booklet points out that Fred is Petra's brother, not boyfriend.) A few years later Peggy got a makeover and morphed into a mini-me of Mattel's popular Francie, "Barbie's MODern cousin" [*sic*]. Petra had her share of accessories, homes, and related play items, went through almost as many identity changes as her predecessor, and finally landed on American toy store shelves in the late 1980s. The most common version has the same side-glancing, molded-lash look of the first Barbie, but the head curiously appears to have been squashed between two fingers (she could be a distant, more affluent relative of Mego's Maddie Mod).

The fascination with Petra for me lies chiefly in her formidable if somewhat musty wardrobe. The sixties/early seventies looks, when compared with Barbie's superbly tailored turnouts, are clearly simplistic and inferior. Yet Petra's penchant for bust darts, three-quarter sleeves, and inverted pleats recalls a forgotten world of fine dressmaker details. Stiff brocade gowns and fur-trimmed coats that reek of old money are endowed with names of European hot spots (Zurich, Milano, Riviera) or heroines from famous operas or Greek mythology (Tosca, Salome, Daphne). The same titles appeared every year, attached to different outfits, until about 1971, confounding today's collectors. Some of these early looks are shameless copies of other doll turnouts; 1964's Paris is really a colorless take on Barbie's famous Solo in the Spotlight, and another set from the same year, Baden-Baden, is nearly identical to Palitoy Tressy's A Walk in the Park, right down to the little dog on a leash.

By the early seventies, a touch of psychedelia creeps in. Cartoonish geometric print tunics, bell-bottomed slacks, and fringed "buckskin" dresses tussle with generic vest-and-skirt sets, bridal gowns, and traditional German dirndls. From about 1975 to 1978, the clothes get a little dizzy. Lines titled Boutique, Festival, or Exclusive are filled with wildly printed floor-length dresses with sharply nipped waists, big tops (capelets, head-swallowing collars, bell sleeves), and even bigger skirts wide enough to knock over underfed Fred with the slightest turn. As if all that was not enough, Petra is further buried under an avalanche of lacy flounces and lace-trimmed floppy hats or shawls (sometimes mistaken for girls' hair bows). In 1978 a newly remodeled Petra and her African-American friend, Donna Sommerwind, finally reach their chic peak with the elegant Haute Couture line. Sleek pantsuits and gowns, a bikini with a monogrammed cover-up, and a standout ethnic print jumpsuit have a breezy sexual allure and suggest the most costly, up-to-the-minute fashions from Europe's premier design houses. Here is an introduction to Petra family dolls and outfits from 1964 to about 1980.

This essay was originally published as an article in the Autumn 2013 issue of *Fashion Doll Quarterly* and is used with permission from FDQ Media.

This fitted dress with an attached capelet, titled Baden-Baden, shamelessly copies A Walk in the Park, a Palitoy Tressy look from about the same year. Both usually pair white with a dark green print; the Bauhaus-style pattern shown here is a rare variation.

An oversized animal print suit from the same year is called Zurich.

An early Petra booklet employs sleek, minimal Bauhaus-like graphic design.

This mid-sixties Sandra black cocktail dress, found in Paris, is an obvious clone of Nizza (Nice), a Petra look from about the same year. The packaging mimics early Petra graphics and even duplicates some Plasty shots of Petra modeling fashions on the back of the box.

Chapter Two: Barbie-size Junk   23

Here is a section from an early-seventies Petra booklet. Three dolls displaying gowns from the Boutique line are presented like full-size mannequins in shop windows (perhaps on the Bahnhofstrasse?) under red awnings.

Two tailored daytime looks in elegant green, red, and gold packaging from about 1969.

An early-seventies redheaded Petra in her original box, with views of the side and back.

Petra, Fred, and Peggy outfits came in small square boxes from 1964 to the end of the seventies. (The boxes had no window cello; the inner cardboard was wrapped in plastic.) Here is a selection of Petra packages from 1969 to the late '70s, grouped together to showcase the many charming graphic motifs used for the boxes. A buoyant green and orange plaid swing coat with thick white buttons, pretty-in-pink pajamas, and flowered bells—what's not to love?

Chapter Two: Barbie-size Junk

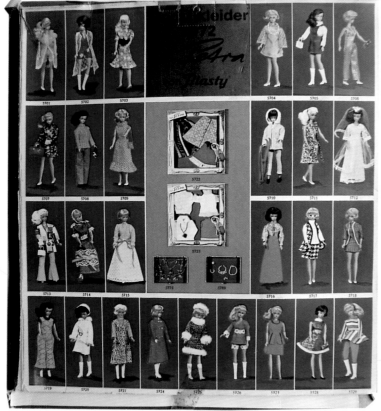

Chapter Two: Barbie-size Junk 27

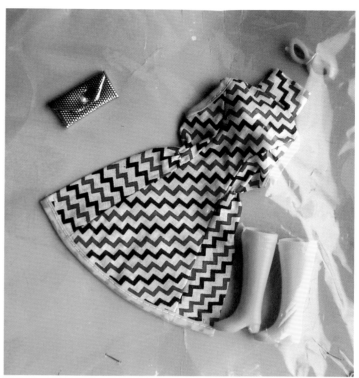

Chapter Two: Barbie-size Junk

Britta was apparently available only as a bride, and is basically Petra with a shorter hairstyle. This version uses an earlier face mold. The boxes for this doll and Petra Star came with clear plastic covers printed with the dolls' names (unfortunately missing here).

A blond Petra Star doll from 1973. Check out the charming seated ladies inside the box.

A lackluster imitation of Barbie's 1967 Caribbean Cruise benefits from Mod-era packaging. And the back is even better than the front!

The Gypsy-Look series of patchwork dresses and separates aimed for a new, younger mood. Five different looks were available.

Chapter Two: Barbie-size Junk 31

A clone of a clone? Jeany is a Petra look-alike with an extra wig, dress, shoes, and a stand.

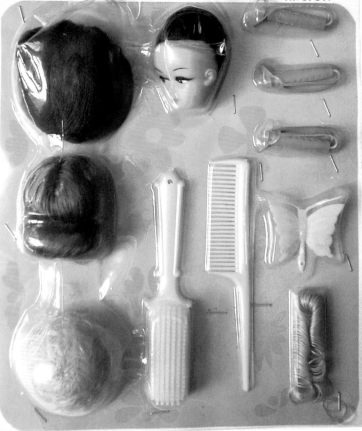

Many different head-plus-wig sets came out in the seventies. How about that cute little barber of Seville?

Every few months a new mystery outfit shows up on eBay. The group in white boxes shown here is probably just the tip of the iceberg.

Petra Boutique outfits in decorative, slightly larger boxes from the mid-seventies. The black box graphics mix fashion figures in period dress with Pennsylvania Dutch–inspired motifs.

Chapter Two: Barbie-size Junk 33

A few more elaborate Boutique sets (one without its box). The gold boxes pair bud vases with sparkling jewelry trees.

An uncatalogued Petra outfit, placed in a clear plastic carrying bag, available for a special reduced price.

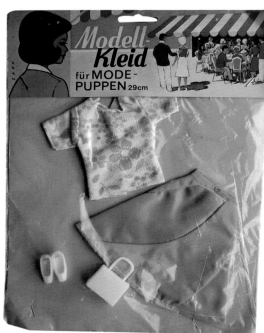

Four repackaged Petra outfits from the early to mid-seventies, with no mention of the Petra name on the header cards.

Chapter Two: Barbie-size Junk   35

Petra separates, usually grouped as blouses, skirts, or pants, came in square boxes.

This shoe and stocking pack reminds me of how cheap Petra packaging is and how difficult it is to find just about any Petra item in its original box. The cardboard used for this set is paper-thin, and the plastic cover is simply stapled onto the card.

On new gold 1976 Boutique Exclusiv cards Petra is photographed like a glamorous, thirties-era MGM starlet, and shares the top header with a vivid pink iris.

Chapter Two: Barbie-size Junk 37

Doll Junk: Collectible and Crazy Fashions from the '70s & '80s

The half-size Boutique cards exchange the iris with a pink rose.

This light blue denim and print pantsuit is typical of early-seventies designer denim—embellished, stiff, and Dry Clean Only.

40 Doll Junk: Collectible and Crazy Fashions from the '70s & '80s

Chapter Two: Barbie-size Junk 41

New late-seventies Petra Boutique Exclusive and smaller Boutique cards invited children to "experience a dream world" with Petra.

Chapter Two: Barbie-size Junk

Chapter Two: Barbie-size Junk 45

Petra embraces prints that Barbie wouldn't touch with a ten-foot (pink) pole. This odd oblong squiggle is one of my favorites. Jonathan Adler, are you there?

Doll Junk: Collectible and Crazy Fashions from the '70s & '80s

Two boxed Petra Princess dolls from about 1978.

Two Petra Sommerwind dolls and a Donna Sommerwind doll, all from about 1978.

Chapter Two: Barbie-size Junk

Petra finally ditches the seventies costume ball, opting for the sleek and sexy 1978 Haute Couture line. These compare favorably with Barbie Superstar and Get Ups 'N Go outfits from the same period.

That's not a pearl necklace you see with the gold-trimmed metallic jumpsuit. As shown on the back of the card, it's a long strand meant to nestle in Petra's luxurious hair.

A hair salon bonnet and chair from about 1978. That's a "smooshed face" Petra shown on the back of the box.

The cardboard Petra Ferienkoffer (holiday suitcase) was a special Airfix promotional item sold in toy stores in the late seventies. Inside the child would find a boxed Petra doll, outfit, and accessory (leftover stock?) to play with.

This delicate rattan lounge chair comes from a Petra Ferienkoffer.

A repackaged Paris Fashion from 1979 or 1980. This patchwork dress is also part of a late-seventies Boutique group of outfits.

Four different Petra dolls from the late seventies to the early eighties. The brunette Petra die Modepuppe from about 1980 appears to share a head mold with earlier Tanya and Karina dolls.

Chapter Two: Barbie-size Junk

The last series of outfits by Airfix displayed the clothes in a gold picture frame. The inner cardboard slid out, revealing a picture for a child to color. This plaid dress and shawl is called "Hütenabend" (evening at a Bavarian hut). The series with blue header cards was not titled.

Horse Riding

Engagement Party

Visit to the Zoo

Grill Party

Flight Holiday

Nearly every Petra fashion collection had at least one traditional German dirndl. This one is called Visit on the Farm.

For me, this 1979 layered look which recalls the Barbie Superstar years is one of those amazing outfits a collector yearns to open and display on a doll. A long white tunic with a brazen V neck and completely open sides slipped over tapered red pants takes shelter under a cropped red and white floral bellhop jacket. What a sexy, sophisticated proportion play!

A Petra hat accessory pack from about 1980.

A late-seventies doll case advertises the Petra Sommerwind doll, four different fashion groups, and features the latest Petra tagline ("Experience a dream world with Petra!").

Chapter Two: Barbie-size Junk   51

The late-seventies Summertime Girls series of dolls and outfits was made by the English toy company Airfix and was probably available only in England (sixty percent of Plasty was sold to Airfix in the mid-seventies). Dolls with a cheaper-looking head mold wore the Petra Sommerwind dolls' original outfits, and all but one of the six available fashions (the pale blue, two-tier tent dress) were taken from the Petra Haute Couture line.

This painfully thin Fred from about 1977 is actually an earlier version, probably leftover stock that was repackaged (a common Plasty practice). He wears a vivid cabana set and apparently beat out Elizabeth Taylor in *Cleopatra* for the Most Liquid Eyeliner award.

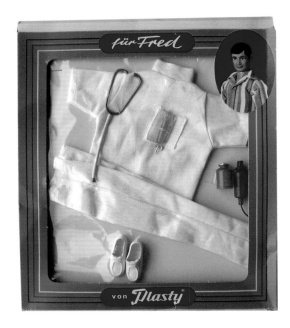

A selection of Fred square box outfits that span the seventies. Two favorites are the green "suede" placket-front overshirt with dippy polyester double-knit pants and the sexy olive green safari suit. The Doctor Kildare set comes with two thermometers (or are they hypodermic needles?) and what looks like an intravenous infusion bottle.

Chapter Two: Barbie-size Junk

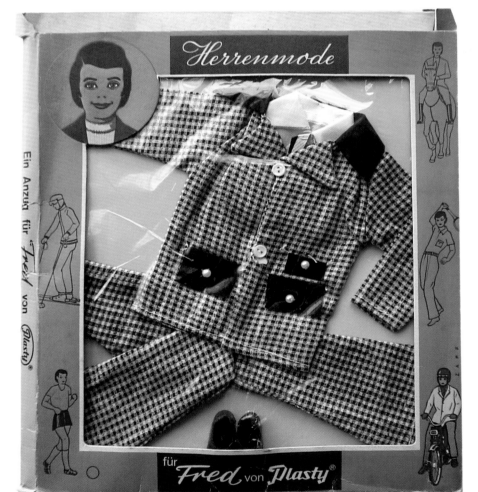

54   Doll Junk: Collectible and Crazy Fashions from the '70s & '80s

Chapter Two: Barbie-size Junk 55

This mid-seventies version of Fred with rooted hair suggests a Keebler elf who has had a sudden growth spurt. He wears a lavender sports jacket set from the second group of Men's Shop fashions.

Two different examples of a groovy Fred shirt-and-slacks set from 1975. The fellow in the upper left corner looks like a young, more effeminate version of Engelbert Humperdinck.

Another mid-seventies Fred with flocked hair and lightweight arms and legs comes in a wood grain–patterned box that matches up with the 1976 Men's Shop fashions packaging.

Hats off to fearless Fred—would Ken have the courage for this futuristic, Cardin-look geometric suit?

Chapter Two: Barbie-size Junk   57

1978 Fred Star sports long rooted hair and easily qualifies as one of the most handsome male fashion dolls EVER. This Fred was available with brown or red hair (both shown, above) and came dressed in a yellow turtleneck and jeans or a copy of the beige leisure suit worn by his American competitor, Now Look Ken. Note the drawing on the side flap showing Fred dressed in a tight ski suit, his goggles nonchalantly worn on his arm, surrounded by women.

On the back of the box, Fred exercises in a second-skin unitard, skis in the Alps, strums a guitar in a jungle paradise (shirtless), and glides across a ballroom dance floor. A girl is always nearby.

Unlike affable and merely well-dressed Ken, Fred Star exudes inbred elegance (despite his copycat Ken leisure suit) and betrays a slightly unnerving, knowing glance.

A few examples from the 1975 and 1976 Fred Men's Shop fashion groups (the first series came in boxes, the second on cards). Some of these high fashion delights, like the furry zip-up jacket, seem light years ahead of Ken's polyester pants duos. Another plus: the wood grain packaging that conjures up the uber-masculine, paneled inner sanctum of a private men's club.

A country club–bound navy blazer and plaid slacks in its original box, and the same set repackaged (a plastic cover is simply glued onto the original cardboard back) at a reduced price.

Chapter Two: Barbie-size Junk

Doll Junk: Collectible and Crazy Fashions from the '70s & '80s

One of the last Fred dolls made by Plasty, from about 1980.

Two versions of a mid-seventies Peggy doll, along with a view of the box back.

A few Peggy outfits from the mid-'60s to about 1979. I bow to the flowered mini-poncho and especially the high-voltage gold bodysuit under a sleeveless orange-red coat with a Mod cap.

Chapter Two: Barbie-size Junk 61

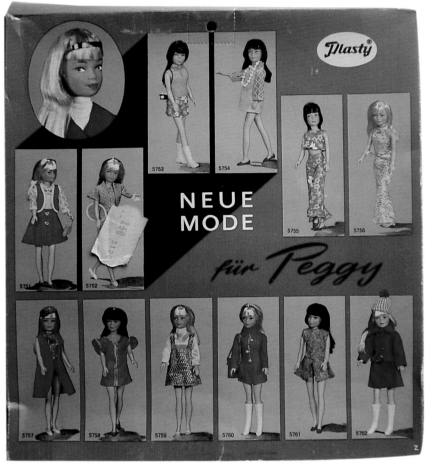

Chapter Two: Barbie-size Junk   63

A Peggy Star from about 1974, dressed in her original short dress and sash. This odd hybrid of Mattel's Skipper and Francie dolls combined the height of the former with the proportions and head mold of the latter.

64   Doll Junk: Collectible and Crazy Fashions from the '70s & '80s

Here's a rare 1975 Plasty booklet that displays fashions for all three Petra family dolls. What at first looks like a mistakenly shrunken Francie next to Petra and Fred is actually Skipper-size Peggy.

Goldköpfchen ("Golden Head") was a 1980 Skipper-size addition to the Petra family, with "growing" hair like the American Character Tressy doll. Her outfits came in the same gold frame packaging that was used for Petra fashions.

MEGO MADDIE MOD

Maddie Mod was my first clone affair and will probably be my last. My allegiance is sometimes off, sometimes on, but it never completely goes away. Will I, or anyone, ever find all the Maddie outfits made from 1967 to about 1978? The early years are well-documented, but the fashions from about 1973 to the end of her run remain elusive. Years of searching on the internet have left me somewhat drained and I am no longer certain of what constitutes an "official" Maddie Mod outfit. Too many fashions are not shown on any package backs, while others turn up on generic cards or repackaged as department store exclusives. Don't even get me started on the endless, arbitrary fabric and pattern substitutions. I have learned to simply go with the flow, and just enjoy the rush when something that COULD be a Maddie comes my way. Her stylish (?) wardrobe reflects both the best and worst of seventies trends. Maddie remains, for me, the most endearing of the fashion clones.

Two Maddies in seventies outfits. The beehive brunette wears a variation of 1970's Cranberry Caper and the blonde with bangs models On Stage from 1972.

Chapter Two: Barbie-size Junk   65

A 1968 Maddie dress titled Black Velvet and two other clone looks in copycat boxes.

A German copy of Maddie's 1970 pantsuit called Green with Envy.

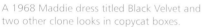

Three Maddie outfits, each from a different collection yet packaged in the same box. Red, White & Blueberry is from 1970, Vanilla Sundae from 1968, and Maddie-dor from a second 1970 group.

This untitled Hot Pants set with a special sticker is from a small group not shown in any known catalog or on any package back.

Doll Junk: Collectible and Crazy Fashions from the '70s & '80s

Some Maddie sets are easily confused with earlier Maddie fashions. I am always getting 1972's Russian Dressing, shown here, mixed up with Eleganza from 1970.

Fringe Binge

Daddy's Girl

Wild Leather

Slim Drome

Rain Supreme

Kool & Kicky

Maddie clothes often turn up without a box, still sewn to their original cardboard liner. Fringe Binge, Rain Supreme, Wild Leather, and Daddy's Girl are from 1970. Slim Drome (?) and Kool & Kicky date from 1972.

Chapter Two: Barbie-size Junk 67

This Montgomery Ward catalog exclusive looks just like a Maddie set sketched on the back of a mid-seventies envelope package. The white dotted outline on the cardboard liner matches up with one from an earlier Maddie box.

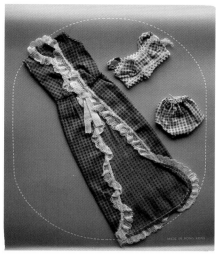

Another Montgomery Ward catalog exclusive, this time similar to Maddie's 1973 Beach Party.

Here is Splash Party, about seven years later, on a nameless Kmart clone. The outfit is also shown on the side flap.

Three 1973 Maddie outfits in new envelope packages: from left to right, Splash Party, Wild Leather, and Scuba Duba. More Maddie madness— this Wild Leather is different from the previously shown Wild Leather, and is shown on a Mego catalog page titled Leather Lovely and on another package back as Leather Is In!

Here is 1973's Blazer's In, also known as Blazer Beauty, in a rare green hanger box. According to the back of the package, the pants should be white. The irregular, hand-drawn window graphics remind me of the "psychedelic" doodles I sketched on my loose-leaf binder back in high school.

68 Doll Junk: Collectible and Crazy Fashions from the '70s & '80s

A rare Midge-like Maddie in her original box from about 1975. This may be the last series of Maddie dolls and fashions to show the Mego logo.

Five nameless looks from the fashion group that accompanied the dolls. Some of these apparently have been sold as Montgomery Ward exclusives as well.

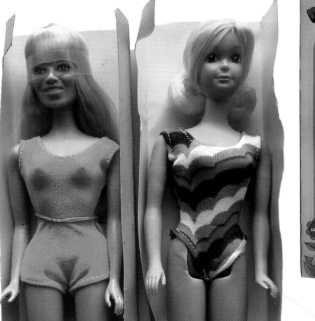

Midge Maddie is shown alongside another Maddie doll from the same series but with a completely different face mold.

Chapter Two: Barbie-size Junk 69

Here are two rare looks from the 1977 envelope package series: a Gunne Sax–like two-piece dress and a patched overalls set, missing a top. The packages say "Summit Industries" and the Mego logo is nowhere to be seen.

This four-piece knit turnout is packaged as a Toys "R" Us exclusive, but is sketched on the back of a late-seventies Maddie outfit card.

Graco Toys cloned Maddie for their late-seventies Miss Fashion World dolls.

This is the third Maddie outfit titled Wild Leather. It's found on the back of the same outfit card. Here, it's packaged as a Lionel Play World exclusive.

70 Doll Junk: Collectible and Crazy Fashions from the '70s & '80s

This hip-tied yellow top and jeans from the Act II group is a duplicate of a Mego Candi fashion from the same year, minus the Candi logo.

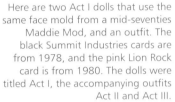

Here are two Act I dolls that use the same face mold from a mid-seventies Maddie Mod, and an outfit. The black Summit Industries cards are from 1978, and the pink Lion Rock card is from 1980. The dolls were titled Act I, the accompanying outfits Act II and Act III.

Two outfits from the same Lion Rock series. The striped jacket duo is Blazer's In, a 1973 Maddie outfit (this time correctly packaged with white pants).

Four more Act II fashions, this time on cardboard figures in baggies.

Chapter Two: Barbie-size Junk   71

The same Act I doll is pictured on a Montgomery Ward box containing a four-outfit set available exclusively through their catalog.

Peggy Ann Doll Clothes, Inc. of Springfield, Massachusetts, made clone outfits for Dawn, Barbie, and larger dolls. Are you digging the Maltese Cross pantsuit? The coral jumpsuit and matching sleeveless coat, too small for 18-inch dolls and too big for Mego celebrity dolls, is a rare set probably meant for the very few 15-inch dolls made in the early seventies (one example would be Uneeda's 1971 Magic Meg).

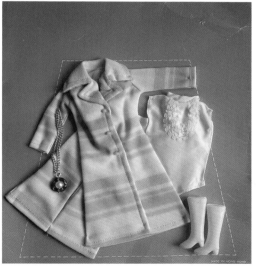

72 Doll Junk: Collectible and Crazy Fashions from the '70s & '80s

A brunette Jennifer wears Fringe Binge, a skimpy red, white, and blue knit hot pants set.

Jennifer is a wide-eyed clone of Mattel's Casey and Twiggy dolls. The logo on her box and fashion booklet looks like it was written by a love-smitten tween (bubble letters, the J and I dotted with hearts). Here is a blonde in a somewhat smashed box.

Jennifer is clearly a budget doll. In her fashion booklet she poses with disproportionate Hot Wheels cars, plastic shrubbery apparently borrowed from a train set, and—shown here—a random toy bull.

Here are some of Jennifer's hard-to-find outfits, without their original boxes but still attached to their inner cards: from top, Hippsie [sic] Gypsy, Long "N" Lovely, Gaucho Gal, Heart Throb, and Tie Power.

Chapter Two: Barbie-size Junk 73

Two Totsy Twistee Hot Pants sets, a sheer gown, and four casual looks from the Dress Down collection (including a German dirndl?).

Bonita A Go-Go has bright orange hair and sienna legs molded with a fishnet pattern. In her garish dress and heavy gold jewelry, she's a Mod version of Joan Crawford in William Castle's delightful schlock shocker, *Strait-Jacket*.

An early-seventies Totsy Cherie stewardess set on an unusual black and white card.

Dynamic Diana, one of countless '70s baggies, put out by Ja-Ru of Jacksonville, Florida.

Doll Junk: Collectible and Crazy Fashions from the '70s & '80s

The Premier Doll Togs Company was based in Brooklyn, New York. The Dodge City suspendered jumpsuit and leopard coat appear on the back of Italian Tanya packages. The patriotic pants duo recalls two Mattel Barbie and Ken sets: Barbie's Sport Star and Ken's Red, White & Wild, both from 1972.

Chapter Two: Barbie-size Junk

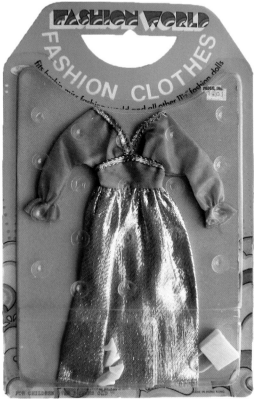

The Larco Fashion World collection also shared looks with the Italian Tanya line. The same photos appear on the packaging of both groups. The red pleather coat with white fur trim could be subtitled "Mrs. Claus goes to Valentino."

LJN's Petite fashions included a smashing Twiggy-esque navy pantsuit with an extended white vinyl yoke and cuffs, flowing palazzos under a pink fur "chubby," a hooded red blanket poncho, and three disco dancers from the later Free N' Easy group. The first two Free N' Easy sets use the same printed fabric found in a rare Mego Cher outfit titled Hanky Panky.

Chapter Two: Barbie-size Junk

Here are Tennis Anyone? (left) and Lazy Days (right), two scarce fashions for Uneeda's 1975 Donna doll. The back of the card shows all twelve looks.

This circa 1970 pink maxi coat from Germany is beyond bland, but check out those three groovy gals on the header card!

A Mini Fashion by Linda from England frames a disappointing version of Yves Saint Laurent's famous 1965 Mondrian dress with four slightly overweight Twiggy-look models in short hair and fishnets.

What movie star would be caught in this shrieking schemata? (Actually, I could easily name one or two, but I won't.) Note the surprisingly sexy red shoes that ape Tuesday Taylor's high-heeled mules.

This wonderful Biba-look maxi in a double-knit Art Nouveau print is a nameless New Fashion from England. The card pays homage to the famous London design house with three unmistakable Biba signatures: butterflies, the model's narrow choker, and deep brown-red lip color.

Three German fashions for Barbie-size dolls from the early seventies. The plaid maxi coat is an excellent copy of Barbie's Madras Mod from 1972.

On a charming German Paris Fashion header card a portrait of a pretty model enjoys a view of the Eiffel Tower from a pitched garret window.

Dirndl, anyone?

I never pass up a big, bold, in-your-face print, and this graphic black and white pattern with intersecting bands of orange is one of my favorites.

Chapter Two: Barbie-size Junk 79

Two European Marion fashions: one from Germany, the other from Australia. The purple bikini top has a gold ring between the two triangles of fabric for the breasts.

Oh, the shame of it! Another German near-exact copy of a Mod-era Barbie outfit—this time, Snug Fuzz.

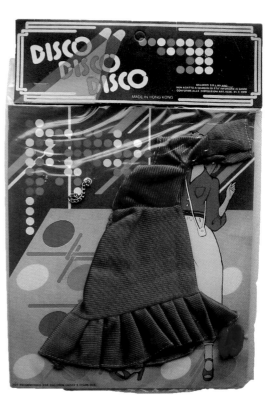

This clever German pantsuit package prints the head of the model on the header and the rest of her body on the cardboard backing.

Two late-seventies disco-worthy fashions. The sectioned metallic mess is from Germany and the purple asymmetric dress comes from Italy.

ITALIAN CEPPI RATTI TANYA DOLL

Tanya, Italy's answer to Barbie, has changed many times over the years. Most fans favor the early dolls, including this one, which shares a face mold with Germany's Karina. Capri Tanya clearly prefers costly status dressing to adolescent pop trends; would today's child take any notice of this adult, somewhat remote doll? She's the picture of reserve and sophistication in a severe chocolate "suede" suit and boots, with an oval Ceppi Ratti wrist tag that simulates stitched leather. She has rooted lashes, the faintest of eyebrows (very seventies) and thick, sideswept hair in an astonishing mix of silver and coppery brown. The box, also brown with silver Art Deco striping, is equally elegant. Except for the modest signature logo, pink is nowhere to be seen.

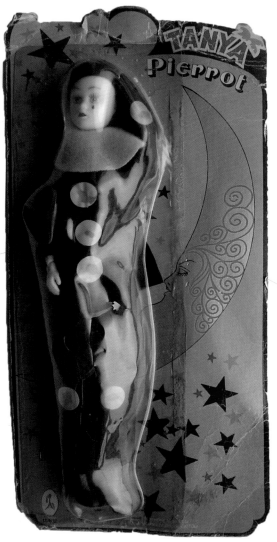

Pierrot Tanya wears the sad clown's traditional black and white costume with oversized buttons and black skull cap. Her face is chalk white and a single tear falls from her left eye. Pierrot's only friend, a female crescent moon, seems to mock her.

The packaging for this version of Tanya on a skateboard (just like Shillman's Adventure Girl) swipes artwork from a 1977 Hasbro Charlie's Angels outfit.

Chapter Two: Barbie-size Junk 81

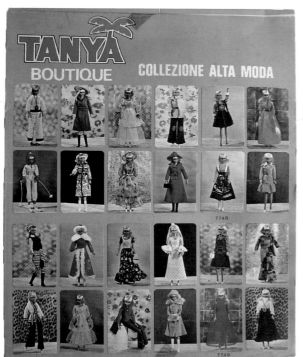

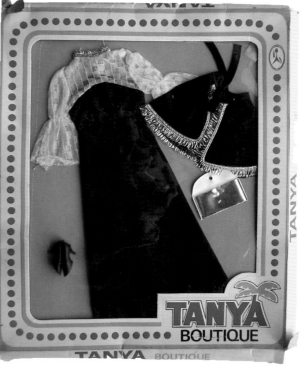

Five circa '72 Tanya outfits from two different series: a newspaper print slicker, a novelty pantsuit, a peacoat and sailor's cap, a camel coat with "leopard" trim, and a pretty empire gown with a tinsel-trimmed capelet. The box backs show what appears to be Maddie Mod dressed in the fashions.

82 Doll Junk: Collectible and Crazy Fashions from the '70s & '80s

A carded Tanya pantsuit and bag. The print is a charming, meaningless mix of emblems, objects, typefaces, and writing (like something you might see on seventies kitchen window curtains).

A mid-seventies tartan-and-turtle trio for Tanya.

A Tanya Intimita lingerie set.

Two gowns from a Tanya bridal collection. The back grid of wedding bells showing the range is pure, smile-inducing corn.

A handsome bag assortment and luggage set, both from the late seventies.

Chapter Two: Barbie-size Junk 83

The Good Nurse Eva of the Nurse Set copies the Mattel Steffie face mold and has sadly lost her legs (some devious shopkeeper apparently cut the plastic bag and stapled the bottom). The title comes from Mattel's 1974 The Sport Set Barbie, Ken, and Yellowstone Kelley dolls.

These two amusing Fashion Flare copies (another misspelled title!) found in Oregon combine the Tanya packaging with a daffy Carol Burnett look-alike.

A double-outfit Wanda/Winny Boutique set tops each fashion with a cardboard head of Wanda or Dawn-size Winny. The larger head resembles Mego's Maddie Mod. The back of the box shows both dolls propped up against the same wallpaper backdrops used for the Tanya shots. At least two fashions appear to copy early-seventies Maddie looks.

Lolita from Germany wears a fetching bob haircut and cutout neckline mini dress.

Super Linna was found in the Netherlands and is a striking, high-quality clone of Mattel's Steffie and Yellowstone Kelley dolls. Her pal Super John can be found in my first book. A range of outfits and a standout wedge-shaped case with a shoulder strap were also available.

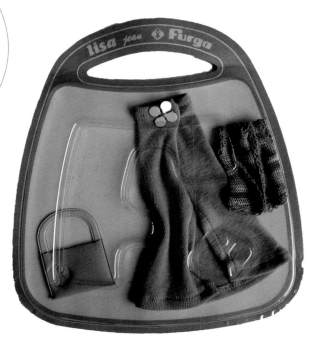

Here are two purse-shaped outfit cards for the Italian Furga Lisa Jean doll and two for her oddly named main squeeze, Boy Jean. Lisa Jean's sweatshirt says "University of Yale." Boy Jean is Ken-size but Lisa Jean has an unfortunate swollen head.

Collette was found in Portugal. With her striking, wide face she looks a little like Madonna.

Chapter Two: Barbie-size Junk   85

SPANISH LILI LEDY BARBARA

Barbara belongs to a small group of Barbie-size dolls made by Spain's Lili Ledy toy company. Barbara was the friend of Senorita Lili, a Tressy clone who first appeared in the late sixties. This version of Barbara from about 1976 has plastered-down bangs (they are actually glued to her forehead) and a slightly heftier appearance than Barbie. She wears her original dress and came with a second outfit. Eleven fashions are shown on her box, each titled after and designed to suggest a global hot spot (a safari suit is called Nairobi, a kimono dress is named Tokio, and so on).

Barbara's friend Ricardo is Ken-size, has molded brown hair, and was available in two different outfits.

This early-seventies music-theme play set from Germany includes a reel-to-reel tape recorder, a portable TV, and miniature versions of real record album covers, including Cat Stevens' *The View From The Top* and *Woyaya* by the British-African band Osibisa (both from about 1971).

1975 MATTEL YOUNG SWEETHEARTS HIS 'N HERS FASHION PAIRS

Mattel hit the nail on the head with these impressive sets. Here are real teenager looks (tasteful and high-end, as if from Saks Fifth Avenue or Lord & Taylor) that might have been worn by The Partridge Family, Donny & Marie, or The Brady Bunch. The box displays five soft-focus illustrations seemingly swiped from a Hallmark card. In each one the same blissfully sexless couple strolls through an idyllic country landscape dressed in an impeccable pairing from the series. Dewy-eyed innocence, fresh flowers, and fashion—who could ask for anything more? P. S.—there is a perfect aural accompaniment to these squishy images; listen to the Monkees' Davy Jones describing "The Day We Fall in Love," if you don't mind a little nausea.

The dolls, Melinda and Michael, don't live up to the perfect world depicted in the package artwork. Doughy Melinda's swollen head rivals that of the Durham Charly doll, and Michael has the hair, but just isn't pretty enough.

DURHAM'S CHARLY

The Durham Charly doll is the quintessential dime store cheapie, a pitiful Barbie-wannabe masquerading as a classy fashion maven with the aid of a precious name and stylish script logo, both cleverly swiped from the early-seventies Revlon fragrance of the same name. (In order to avoid a copyright infringement the name was spelled with a "y" instead of an "ie.") Charly has the round, expressionless face of an unformed child. There is none of the sideways hauteur of Barbie or camp quirkiness of Maddie Mod. The best one can say about the fashions is that they are accurate; they seem devoid of the smile-inducing gaffes in taste found in some clone classics like the Mego Maddie looks. This vacuous non-competitor adamantly, defiantly makes a stand for bland.

Two Charly outfits and a copycat Julie fashion (a Walmart exclusive), all from about 1980.

88 Doll Junk: Collectible and Crazy Fashions from the '70s & '80s

A hard-to-find Charly At Home look, also from about 1980.

The Charly doll sketched on the cover of this case could be a mildly disturbing extra from an animated Tim Burton movie.

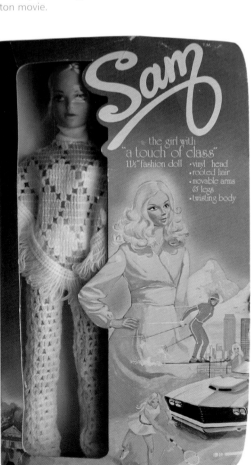

Arco Industries' little-known 1978 Sam, "the girl with a touch of class," might be called a clone of the similarly named Charly doll.

Charly in Fendi-designed fashions?!? Two collections were available exclusively in Europe. The first set came on purple cards, the second set on blue-green cards. Funny how swanky clothes make even the most miserable doll look better.

Chapter Two: Barbie-size Junk   89

A sturdy, nicely detailed luggage and cart set, probably from the mid-seventies, found in Germany.

Two late-seventies Shillman Boutique Fashions copy a fashion sketch swiped from either a New York department store ad or a home sewing catalog.

Betsy Teen and her two fashions were made in the United States but found in Italy.

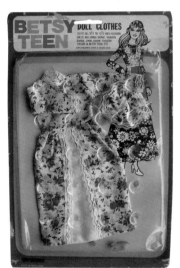

Two scarce Western fashions from the Ertl Wrangler collection.

Here are six Mego-made Montgomery Ward catalog exclusives, repackaged as Smart Set fashions. The furry jacket and pants duo also appears on the back of a Maddie Mod doll box.

I was surprised to see this boxed Kenner Dusty doll in an outfit I did not recognize. She wears a floral peasant blouse and a hopsack skirt with a skinny cord belt. Perhaps she is a department store or catalog exclusive?

Chapter Two: Barbie-size Junk 91

An Olympic Gals track set from Germany. The shorts and hat are labeled "Austria."

Europe's GM Toys offered Superstar Tracey and Dendi Star fashions with packaging that mimicked the Mattel Get Ups 'N Go cards. The orange fur-trimmed jumpsuit is another look from the series.

Clearly the word "Superstar" needs no translation. An outfit card from Italy employs a drawing of a languid lady in a sexy evening print, a single soundbite, and nothing else. Too bad the lumpy winter set doesn't match the promise of the sketch.

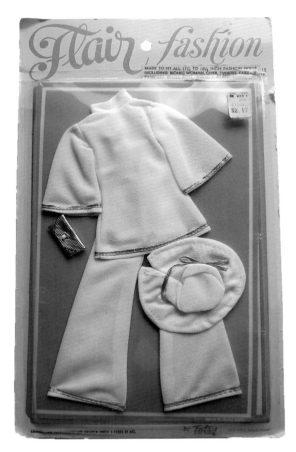

TOTSY FLAIR FASHIONS

The card mentions both Barbie-size dolls and 12½-inch celebrity figures such as Cher, Farrah Fawcett, Bionic Woman and Tennille (minus the Toni), but these two stylish day looks are clearly meant for the larger dolls.

TOTSY FLASHY FASHIONS FOR HIM AND HER

This series, which I only touched on in my first book, was surely inspired by the Mattel Donny & Marie Osmond dolls, reflecting a late-seventies interest in popular television variety show characters that included Sonny & Cher, Farrah Fawcett, The Captain & Tennille, and Diana Ross. Each flashy set for him had a corresponding look for her. Without a Totsy buyer's catalog from this period I have no way of knowing how many outfits were in the series (the backs of the cards are blank). What you see here represents about ten years of searching.

Chapter Two: Barbie-size Junk

This Flashy Fashions set is missing a doll that came with the clothes.

KAYCEE T.V. STAR FASHIONS

These sparkly sets ("Glamorous Fashions to turn your 11½-inch Fashion Doll into a Star") were probably available exclusively at Lionel Kiddie City and Play World toy stores, and no doubt appeared soon after Mattel released the TV's Star Women Fashions for its Cheryl Ladd, Kate Jackson, and Kitty O'Neil dolls. These outfits were titled, just like the Mattel sets, and the six fashions are listed and described on the package back. The two shown here are Lets Dance and On Broadway. The black and red color-blocked gown sketched on the front is titled Premiere Night, and has been spotted repackaged as an LJN fashion (I suspect these were made by Mego).

LJN PETITE SHOW BIZ! FASHIONS

This scarce series of "fabulous Hollywood fashions" for Barbie-size dolls showcases six titled looks: Spanish Gypsy (shown here), Indian Princess, Hollywood Premiere, Cabaret, On Broadway, and Arabian Nights. The last three are also included on the back of the Kaycee T.V. Star Fashions card. This impressive, beautifully colored peasant set layers a fringed purple shawl over no less than three coordinating prints. TV's Rhoda would have loved this one.

Chapter Two: Barbie-size Junk   95

PLAY WORLD TV SHOW STAR FASHIONS

These Mego-made sets were meant for 12½-inch celebrity dolls such as Bionic Woman, Cher, Farrah, Diana Ross, etc. This skinny metallic jumpsuit with a single leg-o'-mutton sleeve was repackaged as a Montgomery Ward catalog exclusive (part of a four-outfit set) for the Mego Cher doll. I am guessing other Ward "exclusives" found their way into these packages as well.

Shillman tried to climb onto the Mego/Cher bandwagon with a glitzy mid-seventies Show Biz line of 12½-inch styles meant specifically for Mego's taller-than-Barbie celebrity dolls. These may have been department store exclusives. Two different styles of packaging are shown.

This clone case with no indication of a manufacturer was obviously intended for Cher. It was available in at least one other color combination.

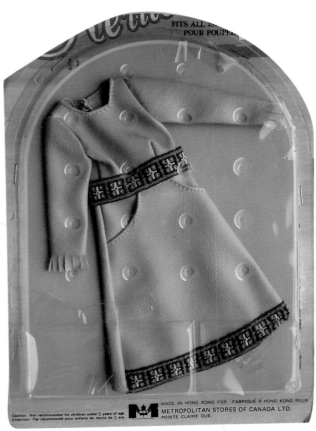

Here is Cher's Mego-made Cherokee, repackaged for the Canadian market.

This wonderful Cher-size Carmen Miranda number was found in Italy.

Chapter Two: Barbie-size Junk   97

Shillman's late-seventies Adventure Girl is a Farrah-fashioned skateboard champion in a silvery shag hairdo and fringed denim shorts. This English/French version of the doll was available exclusively at Zellers department stores in Canada. I have included three increasingly tough-to-find outfits not shown in my first book: Playtime Fun, Shark Fight, and Karate Champ. (Some of these sets were repackaged as Dyna Girl outfits for the European market.)

Caroline, "the doll with... so much more," is a rare brunette Barbie clone who, according to the box back, had two pals, Margaret and Suzan (?).

Sweet Maria is advertised as a Disco Queen, but where's the flash? Where's the sizzle? The back of an outfit package shows no less than six Mego Cher fashions, some apparently altered to fit this 11½-inch doll.

1980 MEGO FASHION CANDI DOLL

Candi represents Mego's last, most ambitious, and I think most interesting attempt to nudge Barbie from the top of the fashion doll heap. The doll's gimmick was a heart-shaped compact that contained 3 colors (red, brown and black) that could be used to dye Candi's hair. The 19-inch version of the doll (which some collectors believe hit toy store shelves first) came with both hair and makeup accessories that included a novel stencil-like "mask" for coloring in the eyelids and lips. But it's Candi's face mold, rather than her fondness for multiple makeups or hair colors, that I find intriguing.

Her look falls somewhere between the striking realism of Superstar Barbie and the almond-eyed stylization of Germany's late-seventies Petra dolls. Despite her glittery getups, Superstar Barbie is the picture of glowing health and sexy athleticism with her clear, direct gaze, teeth-baring grin and active bent arms. At the other end of the spectrum, Petra and Donna Sommerwind, cartoonish, limp sylphs with cool, impenetrable faces, reinforce the tantalizing mystique of the aloof, forever mute fashion model. Candi, not belonging fully to either camp, is more difficult to pin down. Her subtle, "natural look" makeup replaces the ubiquitous blue eyelid with a soft taupe shade and adds barely discernible bottom lashes. The fixed, slightly downward gaze and faint, knowing smile suggest an authenticity injected with a new psychological depth; without getting carried away, there is a sense of introspection, even wistfulness. Candi also happens to look like two stunning actresses: Jacqueline Bisset and, ironically, the now forgotten star of the sixties television series *The Farmer's Daughter*, Inger Stevens, who tragically took her own life in 1970.

Candi had just about everything a top-drawer doll could ask for. A 1980 booklet displays large photos of a French Provincial all-white bedroom set (complete with working bedside table lamp, and breakfast tray), a shower, washstand, dining room set, full kitchen (stocked refrigerator, sink, range, and washing machine), and a novel "Sounds O' Cookin'" electronic oven that mimicked the aural sensations of eggs frying, potatoes boiling, a tea kettle whistling, and a pressure cooker steaming. She lived in a three-story townhouse with a working elevator

For some reason the Canadian market changed Candi's name to Brandi.

(just like Barbie) and relaxed in an above-ground pool with a Jacuzzi and underwater light (an attached cardboard backdrop showed extensive grounds, including large slate stones and brick steps leading to the pool, as well as the requisite patio umbrella, table, and chairs). Her wardrobe accurately echoed late-seventies trends, replacing wasp-waisted ball gowns with roomy "satin" or oriental evening jackets over skinny pants.

In 1981 Mego added a Coppertone Candi doused with that brand's tanning oil scent and a licensed, nameless Jordache doll dressed in designer jeans who shared Candi's face mold. (A mustached male Jordache doll used the Mego Superman face mold.) Outfits that were supposedly copied from real Jordache fashions had a prominent side seam label and came on signature brown cards or in envelope-style packages. Soon leftover Candi clothes and even Cher, Diana Ross, Toni Tennille, and Farrah sets, still stitched to their original cards, found their way into these same brown envelopes. Mego was sounding its first death knell and closed just two years later.

The cover of a 1980 booklet showcases Candi's haunting face.

Chapter Two: Barbie-size Junk 101

Twelve Candi fashions were available on pink cards, including looks for disco dancing and sleepwear.

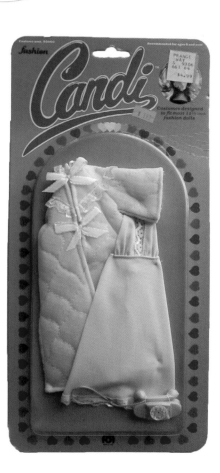

102 Doll Junk: Collectible and Crazy Fashions from the '70s & '80s

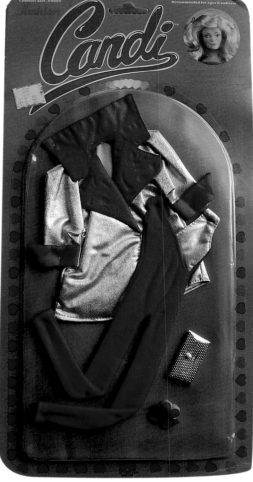

According to a 1980 booklet, the Deluxe assortment of designer originals packaged on purple cards "just might put Candi on the 'Best Dressed' list."

Chapter Two: Barbie-size Junk   103

Sweater Girl

Mego carelessly titled some outfits from Candi's last fashion collection with the same names they used for Cher and Toni Tennille sets. Shown are Sweater Girl (also a Cher look), two versions of Sweet Harmony (also a TT look), Short Stuff, and Charisma.

Short Stuff

Charisma

Here is Sunshine (another TT title), repackaged in a brown Jordache envelope.

104 Doll Junk: Collectible and Crazy Fashions from the '70s & '80s

Australia's Walk Girl comes with a Sony Walkman–style tape player, headphones, and roller skates. Someone cobbled together a confusing card with bits and pieces stolen from other doll packages (the phrase "With Her Tournament It Really Works!!" was taken from an Adventure Girl package and is missing the word "skateboard").

Near the end of Mego's run, leftover, repackaged stock turned up everywhere. This brunette Candi, retitled Deana, was found in England.

1982–83 TOY TIME MISS SERGIO VALENTE FASHION DOLLS AND OUTFITS

A layman could easily mistake any Sergio Valente doll for a Barbie; they are that good. The clothes are equally impressive. Fashions were divided into at least six categories: "Boutique," "Elegant" (gowns), and "Fabulas Furs" (that's not a typo) were packaged in boxes, while the "Designer," "Fashion," and "Activewear" groups were displayed on cards.

The same twenty-six outfits are shown on all the package backs I have seen (these appear to be mostly from the "Designer" and "Elegant" groups, along with eight male fashions), except for the later activewear series, which displays three tops and three bottoms that can be mixed and matched for a total of nine different looks. (Interestingly, these are photographed on a human being, not a doll.) This is unfortunate, as each series had at least six or eight different fashions, some of which are not shown and are therefore unknown to many collectors. To make matters worse, none are titled or even numbered.

I've tried my best here, but in general the Valente fashions are scarce, and show up mint-in-box only sporadically on Ebay.

An African-American Sergio Valente doll models a pale toffee rain jacket with channel-quilted rectangular panels and a dyed-to-match "fur" collar that accurately reflects the early eighties' preoccupation with exaggerated, architectural design. It's from the "Fabulas [sic] Furs" group.

A summer gown with a head-encircling ruffle and sheer overlay skirt. There's a touch of early Madonna in the lace glovettes. The women's outfit packaging displays the slogan "I Love You Sergio!", while the men's packaging uses "Oh, Oh, It's Sergio!" (I can still hear the TV jingle in my head).

This mousy grey Boutique "fur" and "suede" coat with a matching hat and muff has inexplicable green intersecting lines of stitching down the front.

A slender pink metallic gown is swallowed up by a mammoth, floor-sweeping "fur"-edged cape only a drag queen could pull off. A pearl choker, clutch, and shoes are lost in the fluff.

Cozy sweater dressing in ribbed chocolate with clean bands of white. The inserted signature bull's head label is found in just about every outfit.

An impressive miniaturization of a signature Sergio look—a logo tee paired to skinny stitched and labeled jeans. Miss Sergio Valente came dressed in a variation of this outfit. The back and front pockets of the jeans are printed; the stitching on the waistband, down the legs, and at the cuffs is real.

This horsy look pairs a houndstooth/plaid shirt with a breast pocket and a button-down flap to plaid-trimmed moss green trousers.

Chapter Two: Barbie-size Junk   107

Two Western prairie looks from 1983.

This oversized, military-theme white top and skinny pants, cinched in with a wide belt, could have been designed by famous French designers Claude Montana or Thierry Mugler.

Stylish sweats, also from 1983. The package back shows a model wearing life-size versions of the outfits.

Doll Junk: Collectible and Crazy Fashions from the '70s & '80s

Less expensive sets were mounted to narrow cards. A gold-trimmed disco dress and a white pin-tucked top with roomy black pants and beaded belt are two standouts in the group.

A Valente separates set and recolored twin, dropped into a Toys "R" Us Budget baggie.

Chapter Two: Barbie-size Junk 109

When an eighties diva does pants, she doesn't fool around. This 1982 Amy set pairs a mustard glitter jacket with dyed-to-match "satin" zouave pants. A wide green "leather" belt and braided headwrap complete the set. The Amy pairing is shown with seven Sergio Valente sets on the back.

This rare Miss Universe jumpsuit is from 1984.

A 1983 fashion doll trunk displays two glammed-up models, one with a short blond Princess Di haircut.

These three pages from the 1981 Ideal buyer's catalog offer a tantalizing view of dolls and clothes that never found their way to toy store shelves. Barbie-size Laura and her black friend Robin were advertised as "the only fashion dolls that are so soft and real because they are the first dress up dolls not made of hard plastic." The soft-touch theme was reiterated with two Soft Fashion groups: the Soft & Sassy Collection and the Soft & Feminine Collection. Ideal's teeth-baring Lonnie Anderson doll wore a curvy red sleeveless dress and Tuesday Taylor's high-heeled mules. The doll would have come with a full-color 11-by-14 photo of the actress (like LJN's Brooke Shields).

Chapter Two: Barbie-size Junk   111

Two 1986 Show Biz looks from the Toys "R" Us store chain. One copies Cecil Beaton's famous black and white costume for Audrey Hepburn in *My Fair Lady*. The other is a fantasy-theme set possibly inspired by Mattel's Princess of Power dolls and outfits.

This early-eighties white cotton dress, a mini-rendition of an André Courrèges design, was made by Caprice of Paris, France. It captures the designer's playful side with alternating stripes of red and blue and has an "AC" logo in gold near the neck.

A cheapie Galaxy Girl look that cashes in on Mattel's 1985 Barbie Astro fashions.

112   Doll Junk: Collectible and Crazy Fashions from the '70s & '80s

Tonka's space-age Aurora dolls and Fantasy fashions ("The Future Looks Beautiful!") competed with Mattel's mid-eighties Spectra line. Note the clear plastic covers in different shapes.

Chapter Two: Barbie-size Junk   113

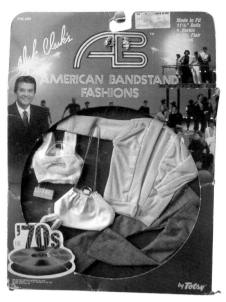

1986 TOTSY DICK CLARK'S AMERICAN BANDSTAND FASHIONS

This series, created to cash in on the revamped American Bandstand television show, probably hit toy store shelves about fifteen years too late. By the mid-eighties MTV and its imitators had eclipsed the decades-long Top 40 Hits program (AB ended in 1989). The box backs are blank so there is no way of knowing how many outfits were in the group. The fuschia metallic '80s set could easily have been packaged as a Rock 'n Roll or Jewel outfit (the top looks too big for Barbie-size dolls).

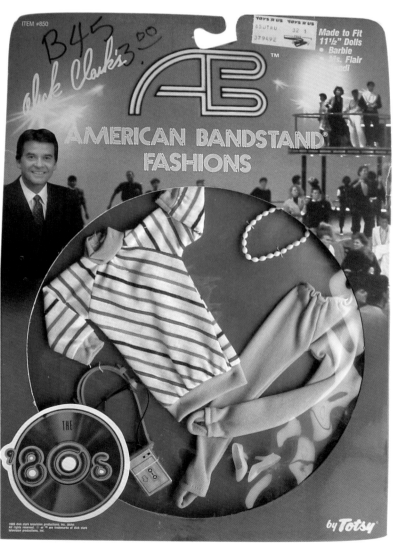

A glow-in-the-dark Toys "R" Us knockoff of Barbie's mid-eighties Dream Glow fashions.

It's hard to believe denim once looked like this. This fitted blue button-front jumper with pink straps was made by Creata.

Chapter Two: Barbie-size Junk 115

The '80s Jem line, undoubtedly one of Hasbro's most successful, influenced countless knockoffs (Mattel beat Hasbro to the punch with Barbie and the Rockers after news of Jem was leaked). Suddenly dolls rocked out on guitars and hand-held organs, had BIG hair in every color of the rainbow, and dressed in tight, cheap, plasticized clothes once thought only appropriate for hookers. The packaging for this Rock 'n Swing Superstar doll has the same dark-to-light color fading seen on Jem boxes and shows two drawings on the back of rocker gals obviously inspired by Jem illustrations.

Shillman's Rock 'n Roll In Concert series included dolls, outfits, and a stage with a working amplifier.

A Superstar Rock 'n Swing fashion has a clear vinyl tote, just like some Barbie Rocker outfits.

1985 BBI TOYS INTERNATIONAL DREAM GIRL USA FASHIONS

The Hollywood Heatwave (best title yet) collection has it all: cropped leotards, tube tops, metallic miniskirts, oversized fishnet hair bows, ruffled boas and more. The black and bright cards take their cue from the Mattel Barbie Rocker boxes.

Chapter Two: Barbie-size Junk 117

These two rocker fashion packages from Italy unfortunately have their tops cut off, but the bottom half of a figure in a Jem-like pose is still discernible at the top right corner.

1986 CREATA ROCK STAR MUSIC-VIDEO FASHIONS FOR 12½-INCH DOLLS

The Creata Rock Star cards blatantly copy the Hasbro Jem graphics and even mention Kimber and Shana, two members of Jem's music group, the Holograms. These sets come with earrings and microphones.

Chapter Two: Barbie-size Junk   119

1986 CREATA LACE FASHIONS WITH FAME FOR 11½-INCH DOLLS

Shocker! A few of the Creata Lace fashions are actually repackaged Rock Star looks. Only the shoes are different (one Lace doll comes dressed in the same "studded" black "leather" dress). The outfits on the pink cards are from the first series. The black cards are from the second series.

By 1987 it wasn't enough to strum a guitar in a tube top and obscenely short skirt. The third and last series of Creata Glow 'N Glitter Lace dolls had glow-in-the-dark makeup, earrings, and spikey hair pieces.

1986 TOTSY JEWEL FASHIONS

These 12½-inch getups were meant for the larger-than-Barbie Hasbro Jem dolls. One or two also turn up in Totsy-made exclusive sets for the JCPenney and Sears catalogs. These fun looks take Jem's "Truly Outrageous" fashions one step further with cheesy fabrics and loopy details that even the Misfits, Jem and the Holograms' dastardly rivals, wouldn't touch.

1986 TOTSY ROCK 'N' ROLL FASHIONS

More Jem-style theatrics with extra cheese, this time for 11½-inch dolls. It's easy to confuse these with the Jewel looks; a similar version of the gold dress with a jagged red apron is part of a rare 20-piece Jewel gift set, and the same music-themed accessories can be found in both lines. A seasoned collector has assured me that a number of these fashions fit both Barbie and Jem.

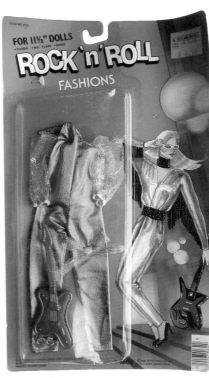

1986 MERITUS SPOTLIGHT ROCK STAR FASHIONS

Jem-sized fashions teamed with a plastic bin filled with generic doll accessories, most of which are too small for Jem (there's a tiny guitar in there that only Dawn could use!). Are these Hasbro-made irregulars or rejects that were never used for the Jem line? The dotted net skirt looks very similar to one found in a Hasbro Jem outfit.

Gloria from Mexico is Barbie-size but with a slightly larger head. She's got a tinge of pink in her hair and is the only rocker doll I've seen that holds a microphone with a long cord attached.

An El Greco Bibi-bo Rockers box from Greece mimics the package graphics from the first series of Barbie and the Rockers. The box back shows the four members of the group: Madonna, Boss, Bibi-bo, and Sandra.

Chapter Two: Barbie-size Junk   123

The Dolly Dots were a Dutch girl band with many catchy pop hits. The Euro version of The Go-Gos had their own doll line in 1984 (made by the German Plasty toy company). Here is a Boy George–baggy, multi-layer outfit complete with rolled-down anklets, head scarf and glovettes—and one member of the group, still attached to her original box liner.

CANADIAN WELLCO SAN/SARRAH ON STAGE

This remarkable Barbie-size doll has to be the most blatant wholesale duplicate of Jem I have ever seen. Wellco didn't simply ape the graphics and illustrations from the Jem packaging. San/Sarrah's dual identity, with separate "day" and "night" personas, is fully explained on the back of the box. Even the words "truly outrageous" are there! The doll has a full head of upswept lavender hair but a somewhat pallid face. She wears a near-exact replica of the Hasbro Roxy doll's original outfit (Roxy was a member of the Holograms' rival band, the Misfits). Like the drawings on the side flap, San/Sarrah is second-rate yet strangely compelling.

This Sarrah set drops the dual identity theme, retaining only half of the original logo. A head shot shows the same doll with "normal" hair and makeup. A V-shape dress with a body-hugging skirt goes over lacy leggings (not tights) and is accessorized with a pink model's portfolio.

1985 MULTI TOYS CORP. L.A. GEAR HOLLYWOOD VIDEO MAGIC FASHIONS

These glitzy, MTV-ready looks came with microphones and/or musical instruments. Gotta love the five punked-out gals in leather, lace, and Lurex at the top of the box.

Chapter Two: Barbie-size Junk 125

1986 MULTI TOYS CORP. L.A. GEAR FLASH FASHIONS

Each tee shirt–style dress in this series had flashing lights that were activated by pressing an arrow on the back. There are some neat '80s New Wave graphics here, but my favorite is the updated take on Roy Lichtenstein's famous romance comics paintings ("Let Him Buy His Own L.A. Gear Wear!"). These look big enough to fit Jem dolls.

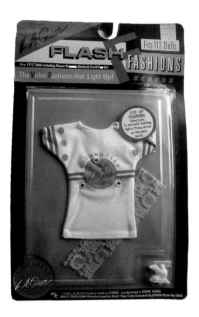

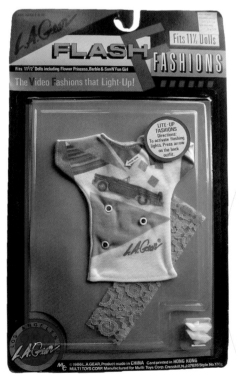

This L.A. Gear aerobic series, Physical Fashions, surely takes its title from the 1981 Olivia Newton-John hit "Physical."

GERMAN BUSCH KARINA

Germany's second-best-known fashion doll, introduced in 1979, is really another version of Italy's late-seventies Tanya with a far more extensive wardrobe. Karina has the same slightly pouty, childlike face as Tanya and the same gesturing hands that are jointed at the wrists, but Busch gave Karina the full star treatment with no less than five different fashion lines (Classic, Top-Modell, City-Chic, International, and Exquisit) and a family of sorts that included a Skipper-like cohort named Topsi and boyfriend named Marc. (Busch was intentionally fuzzy about the identity of these two. An eighties booklet states that Topsi could be Karina's "friend, sister or child," and Marc a "friend, dad or uncle.")

Karina had well-made, though not always exciting, outfits for every possible '80s activity: disco dancing, sunning, tennis, aerobics, and even ping-pong. Early booklets were made to look like miniature fashion magazines; a 1981 cover showing a noticeably tanned Karina in a pink cardigan advertises "89 fashions, chic and not at all expensive," as well as "new for leisure, disco and festive occasions" and "glamorous wedding gowns." Inside the fashion journal was a coupon the child could mail in for a new booklet displaying looks for the next season. Today Karina and Petra collectors appear to be multiplying, and mint-in-box items are not an easy find.

Chapter Two: Barbie-size Junk 127

Chapter Two: Barbie-size Junk 129

A finely detailed accessory set with an oversized pink wallet and miniature German fashion magazine.

An eighties sportswear set repackaged in a nineties baggie.

A late-eighties summer vacation set includes a turquoise suitcase, a matching makeup case with mirror, a pearl bracelet, a passport, and plenty of Deutsche marks.

130 Doll Junk: Collectible and Crazy Fashions from the '70s & '80s

Two scarce Marc fashions from the mid-eighties.

The first version of Marc had molded light or dark brown hair and a penetrating stare. He looks a little like the late actor Steve Forrest.

Chapter Two: Barbie-size Junk   131

Two later Marc dolls with rooted hair. One is Magic Marc with a light-up crystal ball, the other a prince with a tiara for his princess. This Marc bears a striking resemblance to the actor Richard Hatch (remember the '70s TV series *Battlestar Galactica*?).

Marc was sold with and without a painted-on mustache. These two Marcs model a more recent stars-and-stripes homeboy duo called California and a tennis set with lavender stretch undershorts.

132 Doll Junk: Collectible and Crazy Fashions from the '70s & '80s

Here are six Marc fashions from the early nineties. Marc is probably one of the first fashion dolls to get his own laptop. I also dig the early Beatles album and, of course, the Santa suit.

Chapter Two: Barbie-size Junk   133

Can a doll be TOO adorable? Topsi is all bundled up in a snowsuit and scarf for a day of sledding with her canine companion (called Snowy?). This truly childlike doll has a round, pudgy face and button nose, making her a bit more approachable than serene, perfectly proportioned Skipper. Topsi leaves a lump in my throat and a tiny ache in my heart.

Here and on the next page are eight Topsi outfits from about 1984 to 1988. Note that almost all Karina, Marc, and Topsi sets came with a photo inside the box showing the doll dressed in the fashion.

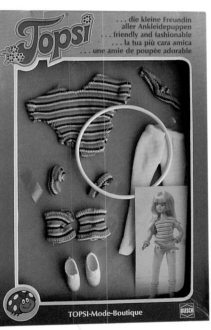

Chapter Two: Barbie-size Junk

This adorable baby named Karinchen comes nestled inside a box not much larger than one meant to hold matches. The doll is a rare promotional handout given to store owners as a thank you for business orders, and was not sold in this packaging. The inside flap gently prods sales with the words "I will not cry at night, but I will sell well during the day. Have you ever ordered me? Daddy Busch is looking forward to your orders."

Three Karina Baby outfits.

SPANISH FAMOSA DARLING

Famosa's Darling doll is more parody than clone, with a frozen expression of surprise and audacious flat feet (Darling wears flats with everything, including ball gowns). Her large wardrobe mixes authentic '80s glamazon getups meant for a boulevard promenade or a glittering fete with a few more restrained Princess Di–type sets. (There are no clothes in this gal's walk-in; every look is an ENSEMBLE, with bags, hats, serapes, and jewelry to match.) Then there is the occasional gaffe that is impossible to explain, like the floor-length red knit dress with a mangy fur neckline and matching Harpo Marx wig. It's easy to make fun of this doll but she's pretty close to irresistible.

This nameless Darling doll resembles the first Barbie and the Rockers doll, yet was released in 1984, a year before the Mattel Rockers.

Golden Nights Darling from the mid-eighties.

Chapter Two: Barbie-size Junk   137

The first series of Darling outfits was not classified. Later outfits were divided into three categories: Diffusion (an industry term for casual, less expensive looks offered by a high-end designer), Boutique (dressy), and Gran Gala (big-time evening).

For me, Vanna White will forever be a goddess of eighties fashion, a lovely mannequin in shoulder pads, sequins, and stilettos. This 1990 Vanna White doll, which came in a number of different dressy looks, was only available through TV's Home Shopping Network.

Authentic, handmade Balinese ceremonial costumes for Barbie and Ken-size dolls, from about the mid-nineties.

Chapter Two: Barbie-size Junk   139

Here are four shameless copycats from Thailand that steal two letters from a recent Mattel Barbie logo. The outfit in the baggie is a rendition of—you guessed it—Ronald McDonald's trademark yellow jumpsuit!

Ken-Size Junk

CHAPTER THREE

1971 MEGO RICHIE DOLL

Most Mego collectors know by now that Fighting Yank and Richie are one and the same, and that Richie clothes represent the glorious apex of early-seventies el cheapo male doll fashion. Here are After Six, Snow King, Tan Trench, and two versions of the African tribal–looking Cool Cardigan. Why Mego changed the name Richie to Ricky and removed their logo from Ricky boxes is anybody's guess. Ricky boxes do not show the full range of outfits made. Over the years I have managed to locate a burgundy double-breasted suit with a white Paul Revere and the Raiders jabot collar shirt, a camel suit with a collarless cardigan jacket, and a metallic tuxedo done in the same fabric as Maddie Mod's 1968 dress, Shimmering Silver.

Chapter Three: Ken-size Junk   141

A black Mego Richie, dressed in Double Dynamite, hangs out with his two German pals, a redheaded Fred Star and Super-Hans, a Ken clone baggie doll.

This clone of the Mego Richie doll was found with the side of his card cut off, so his name and manufacturer remain a mystery. He probably came out in the late seventies and resembles the actor Peter Lupus (the "muscle" from the hit '60s TV show *Mission Impossible*).

Chapter Three: Ken-size Junk 143

I am guessing this Jak Pak Olympic Star was modeled after 1972 swimming sensation Mark Spitz.

Three looks for Totsy's Chuck doll: pajamas and a sponge big enough to wash a car with, a tennis set, and a gray jumpsuit with stitching to simulate pockets.

Two groovy looks for Germany's Ken clone named Toni.

A German Welo "Neue Jeans" outfit demonstrates the early-seventies "unisex" look in fashion with a denim duo sketched on the front of the envelope. On the back the matching duds and background of an appropriated fashion photo have been painted over (I assume to avoid any legal action).

Chapter Three: Ken-size Junk　　145

Two German "Up-To-Date" designs for male dolls. The photo in the top right corner of Talking Ken in Town Turtle is swiped from 1969 and 1970 Mattel Barbie fashion booklets.

Two more male fashion sets from Germany. The sketches on the header cards range from cute and cuddly to cadaverous.

146   Doll Junk: Collectible and Crazy Fashions from the '70s & '80s

Frank knows how to stand out in a crowd.

Three packaged outfits found in Germany but possibly imported from the United States.

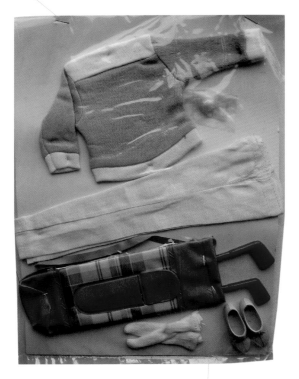

Chapter Three: Ken-size Junk   147

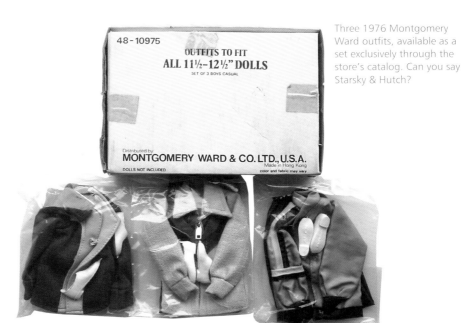

Three 1976 Montgomery Ward outfits, available as a set exclusively through the store's catalog. Can you say Starsky & Hutch?

A pleather Davy Crockett suit, complete with coonskin cap, knife, and belt, from Steha Toys of Germany.

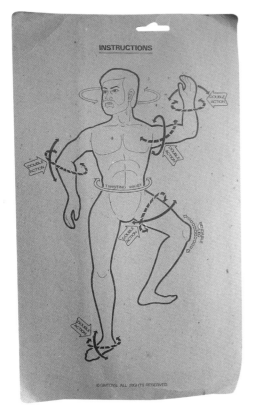

GM Toys' Super Mike is dressed as a fashion doll (couldn't they give the poor guy regular pants instead of light blue leggings?), but his multi-jointed body, grasping hands, and GI Joe hairstyle clearly say "action figure." Super Mike was found on the east coast of the United States.

148   Doll Junk: Collectible and Crazy Fashions from the '70s & '80s

Here are three super-rare Super Mike sets from England. Along with Mike's blinding tuxedo I have included a Mattel Ken tux and variation for comparison.

Chapter Three: Ken-size Junk 149

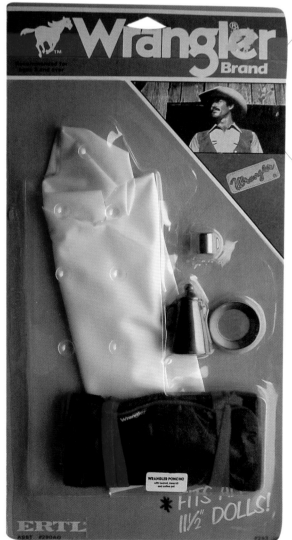

Four looks for the Ertl Wrangler doll: two sets of chaps, a safari suit with a walking stick, a denim jacket with a branding iron, and a poncho with a bedroll and mess kit.

This Ken clone with the exotic name Delon (a French name meaning "bright ray of light") hails from Australia. Note the single oversized arrow at the top of the card.

150   Doll Junk: Collectible and Crazy Fashions from the '70s & '80s

A nameless Mego Jordache doll shares a face mold with the same company's Superman doll.

A Wrangler rodeo accessory set comes with two barrels and a mini Brahma bull.

Chapter Three: Ken-size Junk   151

Five looks from the 1982 Totsy Flair For Him collection. Is that Glen Campbell in the top right corner?!?

152   Doll Junk: Collectible and Crazy Fashions from the '70s & '80s

Three exclusive JCPenney catalog outfits, available as a set, from the early eighties.

A Toys "R" Us exclusive knit sweater and slacks. Check out the mustached Robert Redford clone with noticeable chest hair in the top left corner.

Chapter Three: Ken-size Junk 153

The early-eighties black Sergio Valente doll is identical to the white version, except that he appears to have been literally dipped in chocolate.

Two scarce men's looks: a butterscotch velour outerwear set with a furry pile hood, and an Urban Cowboy outfit with tiny bull heads running down the jean legs and even embossed on the sides of the boots.

These two Toys "R" Us Adventure outfits sewn onto cards are most likely repackaged Sergio Valente sets.

1984 TOTSY SUPER STAR FASHIONS

Some of these MTV-friendly sets, like the many-zippered Beat It jacket, are shameless copies of Michael Jackson's iconic costumes. All of these came with a single glove and most came with high-water pants (two Jackson trademarks). After Jem doll sales hit the stratosphere Totsy slapped confusing yellow stickers on the boxes, stating the fashions could fit 12½-inch dolls such as Jem, Jewel, Rio, and the Misfits (there is no evidence that a Jewel doll was ever made).

Chapter Three: Ken-size Junk 155

156　Doll Junk: Collectible and Crazy Fashions from the '70s & '80s

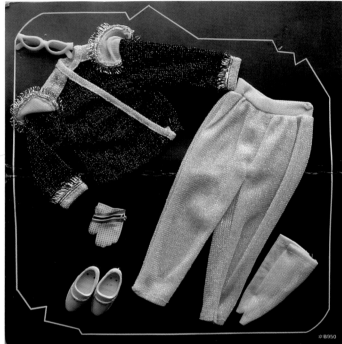

Chapter Three: Ken-size Junk   157

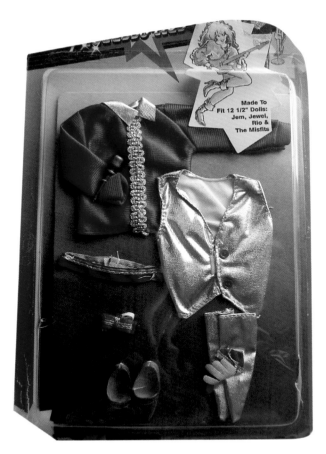

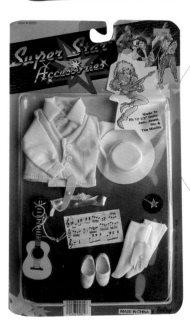

The Super Star Accessory groups are even tougher to find than the fashions. The yellow sweater vest set, shown here two ways, mimics the outfit Jackson wore for the cover sleeve of his 1983 single "Human Nature."

1985 MULTI TOYS CORP. L.A. GEAR HOLLYWOOD VIDEO MAGIC FASHIONS

A second series of fashions for men showcased eight different looks (three more than the women's group). On the back of the card the guys seem to be channeling A Flock of Seagulls.

This ninety-nine-cent Super Stars baggie attaches a Hasbro Rio head to a cheapie hollow body. Other Super Stars hybrids have been found with Hasbro Maxie and Creata dolls' heads.

Chapter Three: Ken-size Junk   159

The El Greco John-John doll from Greece (Bibi-Bo's friend) uses the same face and body mold as Furga's Boy Jean from Italy. In my first book I called the face on the box "enigmatic" and "sexually ambiguous." I now find him disquieting; he looks like trouble.

If this Sun Dreamer fashion doesn't scream "Sunday Golf Outing Dinner at the Country Club circa 1985–Preppy," I don't know what does.

A Creata California Guy purple swagger suit is right out of *Miami Vice*.

This grinning fellow is Prof. Brinkmann, the head doctor from the German television drama series *Die Schwarzwald-Klinik* (The Black Forest Clinic). The popular program ran from 1985 to 1989.

Big Doll Junk

CHAPTER FOUR

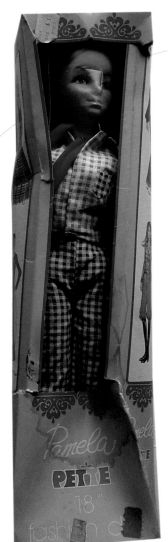

EARLY-SEVENTIES LJN PAMELA

If I could have just one 18-inch fashion doll, it would be LJN's Pamela. Unlike Ideal's Tiffany Taylor, a heavy-lidded glamorpuss who frankly leaves me cold, Pamela seems unaffected and accessible. Her blue-shadowed eyes twinkle and there's a faint smile on her pale pink lips (Tiffany's pretty pout seems permanently frozen, as if she is forever waiting for the next click of the photographer's camera). Pamela comes across as smart and instantly likeable. (You could say she blends Candice Bergen's cool blond looks with Mary Tyler Moore's winning personality.) There are no sizzling siren numbers in this doll's closet, just good clothes that hit on all the seventies trends. Meet Pamela and check out nine looks from her fashion collection: Snowflake, Trail Blazer, Fringed Benefit, High Land Miss (two versions), Green Hornet, Deep Purple, Nifty Norse, and Mucao (or Mucho?) Poncho.

Chapter Four: Big Doll Junk 161

162 Doll Junk: Collectible and Crazy Fashions from the '70s & '80s

A rare Peggy Ann clone set for 18-inch dolls pushes pink polka dots.

Chapter Four: Big Doll Junk

EARLY-SEVENTIES SPANISH VICMA FANNY

If the Spanish Famosa Nancy doll can be compared to Ideal's Crissy, then the Vicma Fanny doll must be placed alongside Ideal's Tiffany Taylor. Both have more slender, teenage proportions and project an older look than either Nancy or Crissy. Yet Fanny is very different from Tiffany. Fanny has a high, "helmet" coiffure and a serenely elegant face that is more realistically modeled, with softer, more subtle makeup. This is a super-sophisticated doll with expensive, highbrow taste. In comparison, the voluptuous Tiffany (even her name pales) appears cartoonishly sexy, and her flashy clothes immediately fade next to Fanny's costly European couture.

Pinacle—A sleek, couture-style take on the pantsuit: an abruptly chopped, sleeveless navy "pull," full matching pants, and a skinny yellow ribbed turtle. The yellow horn-on-a-string only ups the style quotient.

Pampa—Another yummy sportswear look. A "rich hippie" poncho, seemingly cut from a cozy white blanket and trimmed with a wide border of green felt fringe, slips over a navy ribbed turtle and yellow basket-weave pants.

Capucine—The Fanny collection has some of the best examples of seventies-style "separates dressing" I've seen. Here, two handsome prints—a finely detailed, black-ground floral and a graphic red and white "tweed"—conjoin in a button-down shirt and high-waisted bells with matching godet inserts.

Liceo—You knew Fanny had to have at least one fur coat. The white jacket is finger-tip length, and obscures a bare red bias-cut gown with two white "gardenias" just below the plunging neckline.

Cristina—Fanny digs the garden in brown and maize sweater knit overalls with a pale yellow ribbed turtle.

Verbena—Mammoth blue and yellow blooms (almost the size of Fanny's "helmet"-coiffed head) are strewn across a graphic diagonal grid in a floor-length skirt and matching fringed shawl, worn with a simple black tricot turtle.

Saratoga—This striking meteor shower–print gown has a floor-length scarf or panel that is fastened on one side with a small "jeweled" brooch.

Pirineos—A white sweater and navy and white pinstriped flares snuggle under a brown "suede" and white sheared "lamb" jacket, closed with two brown "suede" ties.

Chapter Four: Big Doll Junk 165

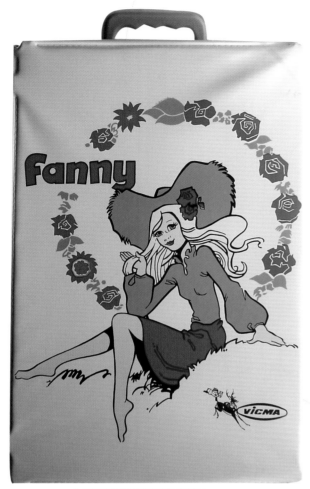

The drawing on this delightful Fanny case makes no attempt to duplicate the appearance of the doll. Instead we get a slightly goofy, barefoot gal who looks as if she weighs all of 80 pounds, lolling on a grassy hill.

A box back from the first series of outfits shows twelve looks. The fashions are modeled by three white Fanny dolls (a blonde, brunette, and redhead) and two black dolls—a brunette and a redhead!

Londres—The following three outfits are from a harder-to-find second series of fashions. Fanny looks very Victorian in an ivory blouse with a cascading jabot neck paired to a simple long brown skirt. A band of embroidery marks the narrow waist.

Cordoba—This half authentic-gaucho-costume, half fashion-adaptation layers a tasseled brown "leather" bolero and chaps over a white blouse and black gaucho pants. Add black boots and a wide-brimmed hat and Fanny is ready to roam the Argentine plains in style.

Mariola—Fanny loves her knits. This time it's a belted empire top with flared pants and a matching cap.

166　Doll Junk: Collectible and Crazy Fashions from the '70s & '80s

Horizontal boxes distinguish the Spanish Jesmar Soraya fashions from the Vicma Fanny looks. Twelve outfits are shown on the back.

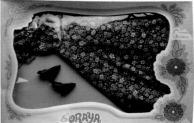

1979–1980 MEGO CANDI SWEET HEART AND DESIGNER SWEETHEART FASHIONS

I have located outfits from what I presume to be three separate groups of Candi fashions Mego put out for their 18-inch doll. Some of these are common while others are extremely rare and may have been made exclusively for the European market.

Two versions of a one-shoulder swimsuit with a hearts-and-flowers, sarong-style skirt.

Candi's first group of Sweet Heart fashions is disappointing. Here is a simple, halter-tied starlet gown in pink.

This second series of eight outfits from 1979, which is not shown on any package back, is both the most stylistically adventurous and most difficult to find. Here, Candi heads for the roller disco in a sporty "wet" purple baseball jacket, gym shorts, and cap, with a white tank.

Candi's sleek jogging suit in immaculate white gets black and red racing stripes down one side of the tank and matching pants. Unfortunately the same open-toe heels accompany every outfit, regardless of the style.

Sporty "Rich Bitch" (RB for short) separates: a glossy mauvey-taupey cropped jacket, a high-collared striped blouse with hidden buttons, and chocolate "leather" pants.

A brilliant gold cami juices up a slim cardigan-style "leather" jacket and matching pants.

Candi makes an elegant "Halstonette" in this no-frills, single-sleeve evening dress detailed with a set-in, curved waistband that echoes the asymmetric hem of the skirt.

Bright "satin" separates: a collared and cuffed baby pink blouse, green tap shorts, and a fringed yellow sash. In this instance the strappy heels lend a vintage thirties look (think of Ruby Keeler tapping away in *Forty-Second Street*).

Candi is a glittering Cher wannabe in this stripes-'n'-skin bolero and petal skirt that use the same patterned metallic knit found in a Bob Mackie-designed Mego Cher gown called Frosted Feathers.

This blue disco jumpsuit with a silver sash was found in England.

A long asymmetrically wrapped dress in light green tricot gets dark green "leather" trim and a wide matching sash.

Chapter Four: Big Doll Junk   169

Another bodysuit/swimsuit-and-skirt duo, this time in green and metallic stripes.

The last series of Designer Sweetheart outfits from 1980 is a little better than the first group. Here's a red, white, and blue sportswear trio in velour, cotton mesh, and denim. Two authentic details: the knotted corners of the jacket hem and the full-cut denim pants with pleats.

Candi boogies in a halter-tied, multi-colored tube top and skinny Lycra leggings. The example at the left has pink string ties at the ankles; the version at the right has white.

Candi's dance recital dress came with a sheer pink skirt (left) or a lavender skirt (right).

A channel-quilted Chinese coolie jacket with blue trim teams with skinny matching pants with slits at the ankles.

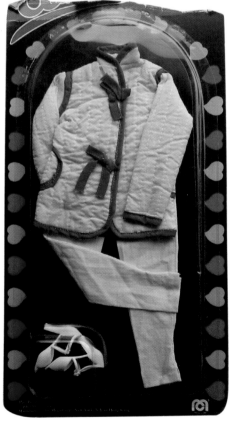

A zippered red terrycloth jumpsuit is seemingly lifted right out of a circa '79 Sears catalog page.

Chapter Four: Big Doll Junk   171

Candi has a new slant on evening finery in a gold-trimmed diagonal-hem dress with white fishnets and a gold pouch bag.

Candi's lone lingerie set pairs a daisy-strewn white "satin" bra or camisole with a matching slip.

This silver and gold jumpsuit (found in England) and silver-splattered variation are not shown on the backs of any Candi fashion cards.

Leftovers

CHAPTER FIVE

The Spanish Famosa Lucas doll (Nancy's friend) has a circa '71 Jane Fonda shag haircut (I'm thinking of *Klute*) and wears, if you can believe it, a belted poplin raincoat. For me, this "little big man" was a must.

Freckles and Freddy, "Uneeda's Olde Fashioned Rag Dolls," exemplify the nostalgia fever that permeated much of the seventies. Here is homespun Freddy from 1973, miraculously untouched in his original box.

Chapter Five: Leftovers   173

Thanks to the colors and clean, simple box graphics, this early-seventies Palitoy Goldilocks "tie dress" and "contrasting knix" packs a real punch. I also dig the purple-tinted photos of the doll on orange rectangles.

Here's more thematic packaging that's hard to resist. Ideal's 1974 Rub-a-Dub Dolly outfits came in adorable blue bathroom tile–printed boxes.

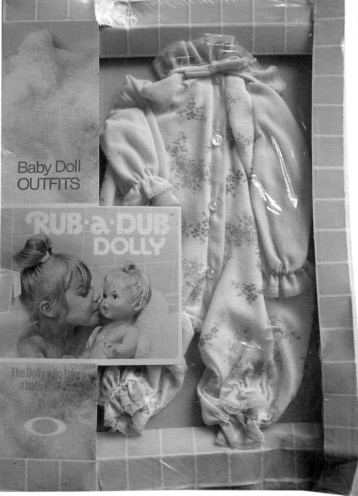

This mid-seventies Famosa Nancy "Mickey Mousse" set from Spain is super-rare and super fun.

Three right-on fashions for larger dolls: a striped fur jacket from the Netherlands and two detailed dresses, one with loopy pockets and the other in an Art Deco circles print, both from Germany.

The Li'L Stuffs Family (I assume Pa, Ma, and Blondie) were cheapo Kmart exclusives that took their cue from both the Mattel Sunshine Family dolls and the TV show *The Waltons*.

Chapter Five: Leftovers

The early-eighties Rainbow Rag Mops, counterfeit cousins of Mattel's beloved Rainbow Brite dolls, are androgynous wonders with button noses, and limp yarn for hair. I don't quite get the angel/devil theme. Here is one of their costumes, Mardi Gras. Eight sets are shown on the back.

Uneeda's 1977 I'm Hanthum doll sucks his thumb when his arm is turned up. This anemic tyke has minimal features and zero personality; besides the thumb gimmick, what is there that could possibly attract a child? He's got the same dead-penny eyes as the rag doll from the Island of Misfit Toys. His female friend, not shown here (why bother?), has blond pigtails and is called I'm Lonethum, which is just sad.

176　Doll Junk: Collectible and Crazy Fashions from the '70s & '80s

This nameless cloth promotional doll, a BMW biker, was made by the Spanish company Paya, a toy firm better known for its finely detailed tin and plastic model cars. She wears a white zippered jumpsuit, carries a red helmet, and clearly needs a better makeup job.

E. Goldberger's 1986 Sugar Pudd'n doll ("Sweeter Than A Chocolate Spoon") could pass for one of the Cosby Kids. This is the only packaging I know of that shows a photo of the doll's creator (in this case James W. Roland) prominently on the front of the box.

Lumina, who literally has stars in her eyes, is sort of frightening (don't look at her for too long before you go to bed). This seldom-seen Rainbow Brite clone made of hard plastic was one of four put out by Fishel Toys.